BLUEPRINT NOTEBOOK

Technical Innovations

This notebook belongs to

ARCHIMEDEAN SCREW

ELECTRIC LAMP

10
04.09.81
FACSIMILE DESIGN

Mar. 23, 1897.

ELECTROPNEUMATIC LOCOMOTIVE

HEAVY DUTY LOCOMOTIVE

RADIO RECEIVER

RADIO TELESCOPE

Aug. 1, 1957
SATELLITE STRUCTURE

TEA BAG

Nov. 10, 1840
PRINTING PRESS

TOY ROBOT

EXCAVATOR

FULLY MOLDED GAS MASK FACEPIECE

Nov. 25, 1986
FIREWORKS

REVOLVING STAIRS

TELEVISION CABINET WITH SLIDING REMOVABLE CHASSIS

COMBINED CASSETTE RECORDER AND RADIO RECEIVER

Aug. 25, 1970
CHAIN SAW

MANEUVERING VALVE FOR HOT AIR BALLOON

TRAFFIC SIGNAL

HELICOPTER

June 16, 1914
TELEPHONE EQUIPMENT

TELEPHONE SET

9d
9c
9
9a
9e
Feb. 20, 1996
8
9b
5a
5d
5c
5b
PORTABLE WIRELESS COMMUNICATION APPARATUS

COMMUNICATIONS DEVICE WITH MOVABLE ELEMENT CONTROL INTERFACE

AUTOMOBILE VEHICLE

AUTOMOBILE

MICROSCOPE WITH SPECIAL LIGHT MODIFIERS

WRITING INSTRUMENT

DIVING SUIT

COFFEE MAKER

METHOD OF AND APPARATUS FOR MAKING DRIP-TYPE COFFEE

POCKET LIGHTER

MANUAL CONTROLLER FOR VIDEO SYSTEM

RADAR AND IFF SYSTEM

BISTATIC RADAR SYSTEM

AIRPLANE

AIRPLANE

Apr. 6, 1982
10
22
20
28
62
BAR CODE CONTROLLED MICROWAVE OVEN

CONSTRUCTION FOR ELECTRONIC OVEN APPLIANCES

Mar. 6, 2001
CONTRAST MEDIUM
COUCH
DETECTOR
HOLDER
X-RAY DIAGNOSTIC SYSTEM

PREFERABLE TO TWO DIMENSIONAL X-RAY DETECTION

PHOTOGRAPHIC CAMERA ACCESSORY

PHOTOGRAPHIC CAMERA

COMPACT DISC DAMPER AND METHOD

ROBOT AMUSEMENT RIDE

MICROSCOPE AND MICROSCOPE SLIDE

VACUUM CLEANER

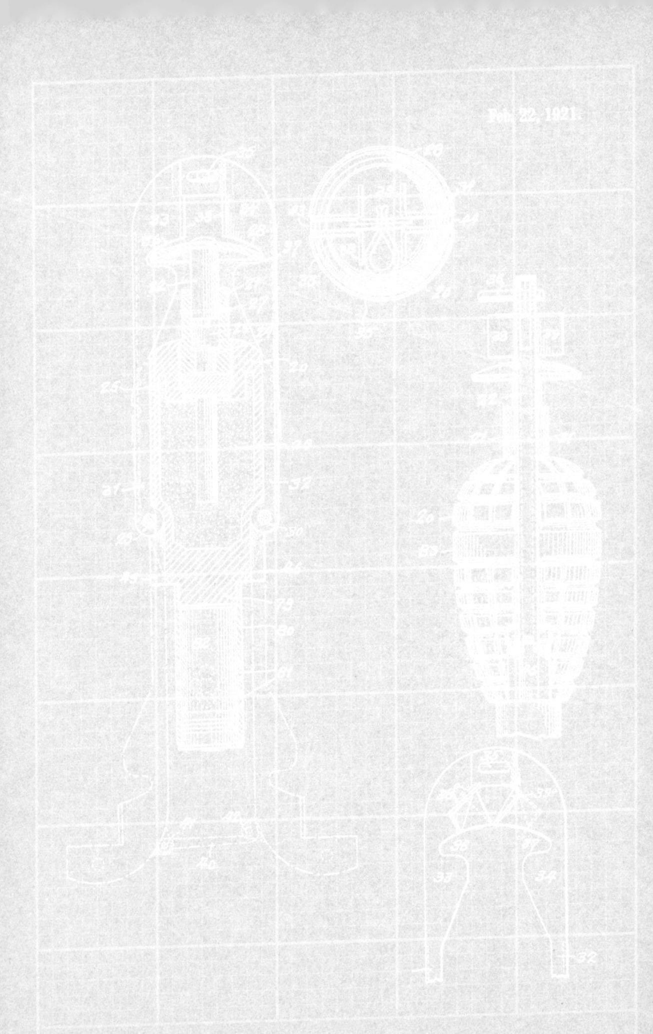

Feb. 22, 1921.
EXPLOSIVE MISSILE

COMBINED WAFFLE BAKER AND GRILL

FIRE TRUCK AND LADDER

FIRE TRUCK

Dec. 11, 1887.)
THRESHING MACHINE

STEAM ENGINE

May 6, 1998
COMPUTER ENCLOSURE

COMPUTER HOUSING

TOASTER WITH WARMING RACK

ELECTRIC TOASTER WITH HEAT UP COOL DOWN BIMETAL TIMER

COMPACT HAND-HELD VIDEO GAME SYSTEM

HAND-HELD TYPE ELECTRONIC GAME HOUSING

Filed April 5, 1961
LIQUIFIER HAVING RESILIENTLY MOUNTED MOTOR AND CONTAINER

CONTROL CIRCUITS FOR ELECTRIC CODING MACHINES

MOBILE SPACE SUIT

SPACE SUIT

SUBMARINE

SUBMARINE

AERIAL CAPSULE EMERGENCY SEPARATION DEVICE

HORIZONTAL TAKEOFF TRANS ATMOSPHERIC LAUNCH SYSTEM

SAFETY RAZOR

AEROSOL SPRAY DEVICE

TOILET BOWL

AIRCRAFT PROPULSION SYSTEM AND POWER UNIT

CASSETTE PLAYER

MULTI-TEMPERATURE REFRIGERATOR

SEPARABLE FASTENER

RECORD PLAYER

DIVER'S WATCH

TWO-ZONE TIMEPIECE

PORTABLE RADIO TRANSMITTING AND RECEIVING SET

PORTABLE TWO WAY RADIO WITH SPLIT UNIVERSAL DEVICE CONNECTOR APPARATUS

300
302
306
Apr. 30, 2013
337
327
331
321
341
311
317
347
370
340
310
345
346
345
343
330
323
333
320
313
NIGHT VISION GOGGLES WITH PELLICLE

TYPE WRITING MACHINE

March 22, 1941
WIND TURBINE

INTERNAL COMBUSTION ENGINE

Blueprint Notebook
Technical Innovations

Printed in Poland
ISBN 978-91-88369-77-2

Dokument Press
Årstavägen 26
120 52 Årsta
Sweden

info@dokument.org
dokument.org